Cornish Inventors

Carolyn Martin

Tor Mark Press • Redruth

The Cornish Tor Mark series

Cover picture: The Trevithick Society's replica of Trevithick's first steam locomotive
Title page: William Bickford

First published 2001 by Tor Mark Press, PO Box 4, Redruth, Cornwall TR16 5YX
© 2001 Carolyn Martin   All rights reserved
ISBN 0-85025-390-X

Author's acknowledgements
I am grateful for the help received from the many libraries and small museums in Cornwall, in particular the Cornish Studies Library at Redruth, and to my husband, as always, for his encouragement and advice.

Photographic acknowledgements
The publishers and authors are grateful to the following for permission to reproduce photographs: Apex Photo Agency Limited, cover; Cornish Studies Library, Redruth, pages 1 and 31; Royal Institution of Cornwall, pages 13, 18 and 24; Teagle Machinery Limited, page 27; WJ Watton, FRPS, page 21.

Printed in Great Britain by R Booth (Troutbeck Press), Mabe, Cornwall

Introduction

For centuries Cornwall has been at the forefront of inventions which have since travelled to all corners of the world. The names of Richard Trevithick and Sir Humphry Davy are well known but there are many others who deserve wider recognition. How many people know that the first house in the world to use gas lighting was in Redruth? William Murdoch invented and installed the system in his house in Cross Street in 1792: the house still stands.

Other significant innovations such as the beginning of the postal service, blood transfusions and the sparking plug all originated in Cornwall.

There is a wealth of information about artists and writers who lived and worked in Cornwall. It is time for the inventors of Cornwall to be better known; after all, they have had a much greater influence on our lives.

In a book this short, omissions are inevitable, notably the mining engineers who need a volume to themselves: only a few representative names have been included. The term inventor is itself open to differing interpretations: is an engineer or designer an inventor in the true sense? Generally engineers have been mentioned if a product or process can be attributed to an individual name, whilst recognizing that every inventor builds on what has gone on before. Teams of designers (such as in aircraft design) do not feature and some unusual and lesser-known inventions are intermingled with the famous names.

Lastly, what is meant by the term 'Cornish'? Here it refers to anyone who was either born in Cornwall or has spent a substantial time living or working in Cornwall.

Ralph Allen 1693-1764

Sir Rowland Hill is generally given the credit for our modern postal system, but it was Ralph Allen from Cornwall who laid the foundations of the present service, over 100 years before the introduction of the Penny Post.

Ralph Allen was brought up at the Duke William Inn at St Blazey. His grandmother was the Postmistress at St Columb

Major; as a boy, Ralph Allen worked with her and acquired a working knowledge of Post Office affairs. Here a Government Inspector noted his aptitude and promoted him to a position at Exeter and later to Bath, where he was made Post Master in 1712 at the age of nineteen. He retained this position until he died in 1764.

When Ralph Allen took over, the postal service as a whole was slow and corrupt. There were six main post roads radiating from London to different parts of the country, including a main route to Falmouth. Postal charges were levied according to the number of sheets and the distance, letters were usually paid on arrival and most were sent to the sorting office in London.

Ralph Allen radically changed procedures and in his contract he paid a fixed yearly rental to the General Post Office, with any excess revenue as his own. The income was considerable, largely because he made the service more efficient and reliable and as the area under his control expanded, the volume of post increased and profits soared.

Besides his postal work, Ralph Allen also owned several quarries at Combe Down, Bath. These produced oolite, the white Bath stone, for building work in the Georgian city. His quarry railway was unique because in order to ensure safety on the steep descent from the quarry to the canal, the wagons had a brake on each wheel – another of his inventions. He worked with the architect John Wood and together they transformed the face of Bath – Ralph Allen was made mayor in 1742. He also owned property at Weymouth, where he invented a special bathing machine for himself.

Ralph Allen devoted his latter years to the development of his elegant classical mansion, Prior Park, Bath, where he entertained well-known political and literary personalities. Henry Fielding made him the model for Squire Allworthy in *Tom Jones*.

John Arnold 1736-1799

A plaque in Arnold's Passage, off Fore Street in Bodmin, marks the site where John Arnold lived and worked, when apprenticed as a watchmaker to his father. A quarrel ended their business part-nership and John Arnold travelled to Holland and then to London, where he was introduced at Court. Here he achieved fame if not fortune and in 1764 he constructed the smallest ever repeating or striking watch, which he set into a ring and presented to George III as a birthday gift.

Next, John Arnold turned his attention to chronometers and invented a new instrument of such quality that Captain James Cook chose to use it on his voyages to the South Seas, 1772-75. Through farming out production and making only the compli-cated parts himself, Arnold practically mass-produced chrono-meters and considerably reduced their cost. In 1779 he created the first pocket chronometer, which so impressed the Astronomer Royal that he decided to test it personally. Arnold's achievement, though, was only recognised by the Board of Longitude after his death when, in 1805, after much legal argument with his rival Earnshaw, his son John Roger Arnold was awarded £3000.

John Arnold's grave is in Chislehurst, Kent; he died in 1799 at the age of 63.

Fernley Hope Banbury 1881-1963

His name certainly does not sound Cornish but Fernley Banbury had a good Cornish pedigree; he was born near St Germans, where his father, Richard Banbury was a farmer and local coun-cillor. He was educated locally before being apprenticed to Bickle & Co, a mining equipment firm in Plymouth. Christine Bickle later became his wife. A turn of fate decided Fernley Banbury's next career move. He was due to leave from Southampton for a post in India, but with late connections, he missed the boat. Instead he started work with a firm of consulting engineers based in Chicago.

In America Banbury made his first acquaintance with the manu-facture of rubber, an expanding industry due to the rising demand for car tyres. He soon realized that there were problems. The mix-ing process was lengthy, labour intensive and above all hazardous and unhealthy because the rubber had to be cut and folded by hand, before being pushed through a type of mangle. Fernley Banbury vowed that he would design a more efficient form of mixer. His revolutionary machine had specially shaped rotors and as it was totally enclosed, carbon contamination was avoided. The first patent was issued in 1916 and the Banbury mixer was adopted worldwide by other manufacturers, for both plastics and rubber.

Despite his success and the many awards that were bestowed during his lifetime, Fernley Banbury never lost his love for Cornwall. He became a naturalized American citizen in 1922 but when he died in 1963 (whilst on holiday in Torquay), his ashes were scattered over his father's grave at St Germans.

William Bickford 1774-1834

Like Sir Humphry Davy's safety lamp, William Bickford's invention of the safety fuse was an attempt to improve the working conditions of the miners.

During the eighteenth century, gunpowder was used for blasting rocks. There was no 'blue touchpaper' and more gunpowder led from the match to the charge, either as an open trail or inserted in hollow goose quills or rushes to slow down the flame. Accidents were frequent, resulting in loss of limbs or blindness.

William Bickford tried to find a safer solution. He experimented by enclosing gunpowder within parchment but, after watching a local rope-maker, he decided to try spinning a trail of gunpowder into a rope. His resulting 'safety rods' were made by impregnating flax yarn with gunpowder and surrounding the yarn with tarred twine to give added strength and waterproofing. In this way, the fire could be conveyed safely to the charge.

The safety rods were first made commercially in 1831 and during that year the factory at Tuckingmill, between Redruth and Camborne, produced 45 miles of fuses. Unfortunately the invention did not gain immediate acceptance: it was expensive and many mine owners continued with the old methods despite the dangers. Gradually it became universally adopted and by 1947 the factory at Tuckingmill produced 300 miles of fuse a day.

William Bickford, the leather merchant from Illogan, did not live to see the success of his invention. He was struck down with paralysis in 1832 and died in 1834. His son-in-law, George Smith, took over the factory and changed the family name to Bickford Smith. After the First World War the business merged with what is now known as ICI and production continued until 1961. All that remains now is a cast tin plaque to the inventor on the wall of the former factory at Tuckingmill.

William Cookworthy 1705-1780

Although William Cookworthy was born in Kingsbridge in Devon, he is included here because he spent much time in Cornwall, both in his search for china clay and as a visiting Quaker preacher. Writing from Plymouth to a friend in 1760 he said, 'I returned from Cornwall two weeks ago, having been there nigh three months.'

William Cookworthy not only discovered china clay at Tregonning Hill, near Germoe in 1746 (and a better quality clay at St Stephen in Brannel some two years later) but in 1768 he took out a patent for 'a kind of porcelain newly invented by me'.

Why was there such an interest in the ingredients for making porcelain? For centuries the Chinese had produced fine, high quality decorated porcelain from petunse (china stone) and kaolin (china clay) using a secret recipe. By the early eighteenth century the Germans had discovered a method of producing porcelain but again this was a closely guarded secret.

China stone and china clay had for many years been used for building work and for lining smelting furnaces in Cornwall but William Cookworthy was the first person to realise its potential in the manufacture of porcelain. In so doing, he laid the foundations for one of Cornwall's most successful industries. Today 60 grades of china clay are produced and 87% of the total production is exported. Modern china clay is largely used as a coating on paper, for paper filling, plastics, paint or in pharmacy – only 12% is used for ceramics.

But to return to William Cookworthy, he was trained as a pharmacist in London and on qualifying at the age of twenty, he set up a pharmacy in Plymouth. In 1768 he established his own New

Invented Porcelain Company in Plymouth, with a branch factory at Truro. Difficulties with production methods, combined with financial problems, forced the closure of the Plymouth Coxside factory in 1770 and the pottery moved to Bristol. Now his porcelain and figures are much sought after as collectors' items.

Henry Crawford 1916-

Many inventions are designed to assist the disabled or elderly and the Level Lift Trolley, designed by Henry Crawford, from St Hilary, is one such example. Originally, the trolley was made to help an elderly friend lift a bucket of coal at the touch of a lever, but it has since been re-designed to raise heavier weights, such as oil drums. The trolley was a winner at the BBC Tomorrow's World International Invention Fair in 1999 and a runner up as the best British invention in 2000. It is now produced by Safety Trolley Systems (STS) in Leedstown near Hayle.

Sir Humphry Davy 1778-1829

Sir Humphry Davy invented the miner's lamp and his statue, standing boldly astride the streets of Penzance, bears witness to his links with the town of his birth.

Sir Humphry Davy was born in Market Jew Street but he spent his early years in nearby Ludgvan. The eldest of five children, he was taught at Penzance Grammar School before being apprenticed to Bingham Borlase, an apothecary in Penzance.

Although Davy had completed his formal education by this stage, he continued to experiment and study science-related subjects. He taught himself French and wrote and published poetry, which was later rated highly by some of the most eminent poets of the day. He was a keen fisherman and wrote about fly-fishing. Even when he reached the realms of fame, he still remembered his mother's marinated pilchards.

His promotion to the Pneumatic Institute in Bristol was the real turning point in his career and although he still held Cornwall foremost in his affections, he rarely returned to the county after moving away. By experimenting he discovered the anaesthetic effects of nitrous oxide (laughing gas) and he was made Director

The statue of Humphry Davy has pride of place in his home town of Penzance.

If you look more closely at the statue, you may be surprised to see that the middle waistcoat button is missing and think that this was an oversight on the part of the sculptor. But no, apparently just a suggestion of his wife's negligence!

of the laboratory in Bristol in 1801 and Assistant Lecturer, then Professor, in Chemistry at the Royal Institution. He was an eloquent speaker and his lectures were packed to overflowing. At the same time he constructed the largest battery ever built and researched tirelessly into the effects of electricity on chemical compounds, isolating and naming seven previously unknown elements. He was knighted in 1811 and in the same year he married Jane Apreece, a rich widow and later (together with his assistant, Michael Faraday) they set out on a grand tour of Europe.

The name of Humphry Davy is generally associated with the miner's safety lamp. At that time illumination underground was by means of a naked flame and this could easily ignite combustible gases, notably methane. It took Humphry Davy just fourteen days to solve the problem in 1815 and, in order to introduce it into the mines more quickly, he declined to take out a patent. He discovered that the heat of the flame would not pass along a narrow metal tube and similarly, by enclosing the flame with a piece of metal gauze, he prevented the flame igniting the gases. The light given out, although dim, was sufficiently bright for the miners to carry out their work. Another less known invention was his suggestion that the corrosion of copper-bottomed ships could be prevented by means of zinc and iron sheathing.

During his later years Davy spent much time travelling and working in Europe. He was increasingly beset by ill health and died in Geneva in 1829.

John Edyvean fl 1780s

Unlike the Midlands and the North, Cornwall did not take part in the great eighteenth century canal boom – apart, that is, from one scheme in North Cornwall, where in 1774, blind John Edyvean, a farmer from St Austell, put forward a proposal to link Bude with the river Tamar. The main purpose of the canal was to transport inland the local sea sand (which contains many crushed mollusc shells) as fertiliser to be scattered on the lime hungry fields. The original scheme covered a direct distance of 28 miles but the circuitous route involved a journey of 90 miles. The terrain was hilly and in order to overcome the problem of the gradients, John Edyvean made his original suggestion for inclined planes on the canal. Trucks would be loaded with the cargo for each descent and ascent, without the use of locks. His idea was developed by others and later examples of inclined planes transported the entire boat up the slope. Three inclined planes were suggested for the Bude end of the canal and two to connect with the Tamar. Unfortunately, although the proposal was approved by Parliament, there were financial difficulties and the scheme was abandoned.

At around the same time (1773), Edyvean suggested a similar scheme for inclined planes on a projected horseshoe shaped canal between Mawgan Porth and St Columb Porth. To quote from *Lake's Parochial History* (1867) 'after proceeding with the work for several years, wasting his own fortune and a considerable portion of a sister's, his scheme was left unfinished for want of capital… This complication of fortunes, it is said, broke his heart.'

Construction work on the Bude canal, as we know it today, began in 1819 and it was open for traffic from 1826 until 1902. Sections have since been re-opened for recreational use.

Robert Were Fox 1789-1877

No book about scientists in Cornwall would be complete without mention of the Fox family, Quakers and shipping merchants from Falmouth. They were instrumental in the establishment of the Royal Cornwall Polytechnic in 1833, to 'stimulate the ingenuity of the young, to promote industrious habits among the working classes and to elicit the inventive powers of the community at large'.

One member of the Fox family was a distinguished inventor in his own right, Robert Were Fox. He experimented with magnetism and electricity and discovered that the internal temperature of the earth increased with depth. His main invention was the dipping needle compass or deflector (1831), an instrument that was particularly useful in deep Cornish mines because the underground readings were no longer distorted by magnetic iron. An improved version of the compass was used by Captain J Ross on his Antarctic survey expedition in 1840-41.

Robert Were Fox's daughter Caroline (1819-71) kept a detailed journal recording conversations and visits to their home by literary and scientific figures, including John Stuart Mill, Carlyle and Tennyson. In addition, the Fox family have left a lasting legacy to the Falmouth area; the Polytechnic continues to flourish and visitors still enjoy the beauty of their historic gardens.

Tony Gilbert 1949-

'Snail post' has a double-edged meaning in Cornwall. Not only is the conventional post slower than e-mail but letters in Cornwall are attacked and even eaten by snails! It seems that they are partial to the adhesive on the stamps and envelopes and often chew their way through whole envelopes and their contents.

The Royal Mail has experimented with several deterrents, fitting a flap to the entrance (the snails just climbed under the flap) or laying copper wires around the opening – but when the copper strips became dirty they lost their natural charge and the snails returned.

Tony Gilbert, a postal engineer came up with a novel solution, now on trial in the Truro area. His snail excluder is quite simple, a row of bristles attached to the upper lip of the posting aperture. The snails dislike the sharp, dry bristles and stay away.

Daniel Gumb ?-1776

High on the moors, not far from the Cheesewring, lies Daniel Gumb's cave, marked by the initials and date 'DG 1735'. Now just a rough shelter, with some rocks inscribed with mathematical symbols, the cave is a shadow of its former glory. When lived in, it had a series of connecting rooms lined with granite, with chairs and beds carved from granite slabs. At the entrance there was a unique granite sliding door, invented and constructed by Daniel Gumb.

However, Gumb's genius extended beyond furnishing his cave: he was a self-taught mathematician and astronomer and, as evidence, Euclid's theorum can still be seen carved into the roof of the cave. Here he brought up his large family and between 1732 and 1743 he is said to have married three times. He was born in Linkinhorne at the beginning of the eighteenth century and lived in the parish for the rest of his life, earning his living as a stonecutter at the local quarry, whilst contemplating the greater wonders of the universe.

Sir Goldsworthy Gurney 1793-1875

'Bude's forgotten genius.' Most scientists are content with just one invention to their name but Goldsworthy Gurney, with his active and restless ingenuity, was responsible for a multiplicity of disparate ideas. He was born in 1793 at Treator, near Padstow and buried at Launcells, near Stratton, at the age of 82. Although he spent much of his life in London, his true home and haven in times of stress was Bude.

Goldsworthy Gurney was educated at Truro Grammar School before being apprenticed as a medical student in Wadebridge, taking over the practice in 1813. Soon afterwards he married Elizabeth Symons, the daughter of a local farmer. Their daughter, Anna Jane, was born in 1815 and she was to remain with her

An artist's impression of Gurney's steam coach, which never ran: passengers were scared of sitting directly above the boiler!

father throughout his life as his companion and advocate.

By 1820, Goldsworthy Gurney felt that he had outlived his time in Cornwall and moved to the stimulating atmosphere of London, where he established himself as a surgeon and was soon assimilated into the leading scientific circles. In 1822 he was appointed as a lecturer in chemistry and natural philosophy at the Surrey Institution – the word scientist was only introduced around this time; prior to this they were known as natural philosophers.

A series of significant inventions followed, most notably his steam carriage, which foreshadowed the motor car. The early models unfortunately did not inspire confidence because the steam boiler was placed within the coach and under the passenger seat. To counteract passenger fears, he later designed a separate vehicle, called the 'Gurney Drag', to pull the main passenger coach. In 1829 the new vehicle was given a road test on a trial run between London and Bath, with much publicity and problems.

The return journey was more successful and was, they claimed, the first long road journey at an average speed of 15 mph. Despite this promising start, opposition came from the railways, the mail-coach owners and the prohibitive turnpike tolls, which spelt financial ruin and the demise of the Gurney Steam Carriage Company.

Undaunted by this setback, Goldsworthy Gurney channelled his energies into other activities. He returned to Bude in the 1830s and built his famous 'Castle on the sand', the first building to be constructed on sand, using a concrete raft. Here he installed 'Bude light' (oxygen introduced into the centre of a flame, to produce a bright light, with reflecting mirrors).

The experiment was extended to lighthouses, where revolving frames gave a flashing beam and individual lighthouses could be identified by the intervals of their flashes. To celebrate the Millennium with Bude light, a nine metre high coloured concrete beacon, with the latest fibre optic techniques, was unveiled outside the castle in the summer of 2000.

Goldsworthy Gurney also worked on a safer method of stage lighting. He created 'limelight' – so it is only possible to be 'in the limelight' due to his invention!

In 1862 he obtained a patent for the Gurney Stove for warming and moistening air, which is still used in many churches today. He was knighted in 1863.

Lemael or Lemon Hart 1768-1845

Lemael Hart, the grandson of a German immigrant who had established a wine and spirit business in Penzance, earned the distinction of being the first supplier of rum to the Royal Navy when he persuaded Admiral Tott to give a measure of rum to every seaman at sundown. Hence the term, 'a tot of rum'.

The distinctive blend of rum invented by Lemael Hart has a mention in *The Oxford Companion to British History* which states: 'The Cornish retain their cultural richness, Britain and the wider world…would be poorer without their cream and Mr Lemon Hart's rum.'

The Holman family

For some, the name Holman is inextricably linked with the Holman Rock Drill (patented in 1881), whereas for others it is affectionately associated with the Cornish cooking range. But there is no doubt that the Holman family have been part of the Cornish industrial scene for generations, with many inventions to their credit.

Nicholas Holman established his boiler works at Pool near Camborne in 1801 and it was here that the boiler for Trevithick's locomotive is said to have been made. But even before this, the family held patents for new methods of raising minerals from mines and anchors from the sea and for patching boilers without letting out the water. During the nineteenth century, the firm expanded with a foundry at St Just and further works at Camborne and Hayle, as well as a dry dock and foundry at Penzance.

Over the years, Holmans have survived recessions in the mining industry by diversifying their products – moving from mining to make pneumatic industrial tools, air compressors and complementary equipment. During both world wars they were heavily involved in munitions work and in 1959 they launched a British first with their Rotair air compressor. Overseas sales have steadily increased and consistently account for 70% of total turnover. The firm, now CompAir UK, is part of the Invensys Group but Holman Rock Drills are still at work in mines all over the world.

Robert Hunt 1807-1887

The name of Robert Hunt, a pioneer photographer, can be seen above the door of the former Robert Hunt Museum, next to the Redruth library in Clinton Road, built in 1891 in recognition of his work in Cornwall. The museum closed in 1953 and the mineral collection was transferred to the Camborne School of Mines Museum.

Although he was born in Devonport, Robert Hunt was intrinsically associated with life in Cornwall; he lived in Penzance and owned property at Fowey and was Secretary of the Royal

Cornwall Polytechnic Society in Falmouth from 1840-45. For the next 37 years he was the Keeper of the Mining Record Office in London and compiled detailed mineral and mining statistics. In his later years he was concerned with the education of miners and established the Miners' Association for Devon and Cornwall in 1859.

Robert Hunt trained in London as a chemist and experimented extensively with light, electricity and heat but his main interest was in photography. In 1843 he discovered the use of iron sulphate (ferrous sulphate) as a developer, an invention that was to prove of great significance in the world of amateur photography. Fox Talbot claimed that gallic acid (for which he held the patent) had similar properties, but this was quite untrue.

As well as his photographic and statistical work, Robert Hunt wrote and lectured with enthusiasm on a wide variety of topics, scientific and literary. As Caroline Fox commented in her Journal on October 5th, 1842, 'Attended Hunt's lecture on chemistry, very pretty, popular, explosive and luminous.'

Sir Henry James 1803-1877

Henry James was born in Cornwall, at Rose in Vale, near St Agnes, but he spent most of his working life away from the county. After military training at Woolwich he was employed in various branches of the Ordnance Survey before being appointed as Director General in 1854. By the time that he retired twenty years later, he had introduced some farsighted processes into the service, most notably photozincography, a term that he invented.

When he took up his new post at Southampton, there were conflicting opinions about the most suitable scale for the ordnance survey maps and the cumbersome pantograph was used for reducing maps from one scale to another. James realized the potential of photography, especially when combined with zincography (printing with a zinc plate rather than an engraved copper plate or stone) and the process was established at the Ordnance Survey in 1859.

In 1860 James was knighted and he went on to build a photographic workshop, using the sun to develop the prints. His other achievements included making the OS publications easier to read and producing maps more quickly and accurately; he made a study of notable archaeological sites and carried out an important survey of Jerusalem.

He co-operated with other European countries to standardise mapping procedures and in 1863 he was appointed Commander of the Royal Order of Isabella the Catholic (of Spain).

Major General Sir Henry James left his mark on the OS office in many ways; he had his initials carved over doorways and erected engraved plaques outside buildings that he claimed to have designed. He has been described as an egotist and self-publicist, more inventive than inventor but was nevertheless a true innovator.

Joel Lean (1779-1856) and Michael Loam (1797-1871)

At the beginning of the nineteenth century, life for Cornish tin miners was hard, with long hours and unhealthy working conditions. Often they had to walk considerable distances to work. Once at the mine, they had the long descent underground and the gruelling ascent after a day's work. Many accidents and deaths occurred, with miners falling off the vertical ladders through tiredness and exhaustion.

The Man Engine was devised to carry men up and down the shaft. The idea for the first effective mechanism is attributed to Joel Lean although Michael Loam, from Gwennap, was the first person to design and construct a working example, at Tresavean Mine, near Redruth in 1842 – in response to a prize offered by the Royal Cornwall Polytechnic Society.

The machine consisted of long rods attached to platforms in the mineshaft, which in turn were linked to the beam engine. As the engine moved up and down, the miners would step between the platforms at the top and bottom of each stroke, moving mechanically up or down the mine. Unfortunately the idea was not widely supported by the mine owners, who thought that installation was too expensive. Following a dramatic accident in 1919, when 31 miners were killed on the man engine at Levant Mine, the system was phased out and replaced by more modern appliances.

The Man Engine at Dolcoath mine, 1893 – see page 17

Alexander Lodge 1881-1938

The Island House at Newquay, set apart and yet so close to the main beaches and town, holds a special fascination. Approached by a footbridge, it was built in 1910 for a Canadian recluse, but since then it has been the home of several eminent personalities in the world of science, notably Alexander Lodge, the inventor of the Lodge brand of sparking plug, and his father Sir Oliver Lodge, the distinguished physicist with an interest in spiritualism and wireless telegraphy. Sir Oliver, together with Alexander Muirhead, is said to have invented the *idea* of radio some years before 1901 when Marconi sent his signals across the Atlantic from Poldhu in Cornwall.

The spark plug (the ignition device in an engine) developed from Sir Oliver's experimental work on electric condenser discharges. Alexander and his brother Brodie took out a patent for an improved system of high tension ignition in 1903 and the following year they formed the firm Lodge Bros. in Birmingham. As Alexander designed new weatherproof plugs, the company expanded, moving production to Rugby and in 1919 the name was changed to Lodge Plugs Ltd. During the First World War the company diversified, making plugs for armoured cars and aeroplanes, but when hostilities ceased they concentrated on the motor car. By the 1930s Lodge was synonymous with spark plugs and the name lives on: an Italian firm has recently re-launched the Gold Lodge spark plug.

Alexander Lodge used the Island House at Newquay and neighbouring Skerryvor as a holiday home. He died in 1938.

Richard Lower 1631-1691 and
Percy Lane Oliver 1878-1944

Richard Lower, who was born at Tremeer, Bodmin, was the first person to experiment with the transfusion of blood from one animal to another and later between humans. His treatise on blood transfusion became a standard textbook and he was physician to Charles II. He died in London but was buried at St Tudy and in his will he left £40 to the poor of two parishes in Cornwall.

Percy Lane Oliver from St Ives was the inspiration behind the establishment of the National Blood Transfusion Service. He was awarded an OBE in 1918 for his work with refugees but he gained even greater recognition for his efforts in organizing panels of volunteer blood donors, who could be called on in times of emergency.

John Loudon Macadam 1756-1836

Richard Trevithick may well have tested his London carriage on Macadam's road because between 1798 and 1802 John Macadam lived at Flushing, where he worked as a navy victualling officer. But his interest lay in road building using broken granite, bound together with gravel and with a camber for drainage.

He built the turnpike road between Truro and Kiggon (Tresillian). Our road surfaces are now 'macadamised', using tar or asphalt, immortalizing the name of our greatest road builder.

William Murdoch 1754-1839

In comparison with Trevithick Day in nearby Camborne, Murdoch Day each June in Redruth is quite a genteel affair, with shop assistants dressed up in lace caps and aprons, and market stalls sprawling up and down the main street. There are processions, concerts, a church service and a commemorative exhibition at Murdoch House in Cross Street, where William Murdoch lived between 1782 and 1798.

Although William Murdock (who later changed his name to Murdoch) was born in Ayrshire, many of his inventions were made in Cornwall. In 1777 he started work at Matthew Boulton's Soho foundry in Birmingham and two years later he was sent to Cornwall as Manager of the firm's Redruth office. Here, he worked as a steam engine erector but his chief responsibility was to check on infringements of the Boulton and Watt steam engine patents – an unenviable task in a hostile environment.

George Stephenson and Richard Trevithick are popularly associated with the steam engine but it was William Murdoch who built the first model steam driven self-propelling locomotive in

A replica of Murdoch's steam Flyer, seen outside his house in Cross Street, Redruth, before being located on the roundabout outside Tesco's

1784, some seven years before Trevithick. He tested it in Church Lane, Redruth and it is said to have terrified the vicar. In 1785, Murdoch married Anne Paynter but sadly she died in 1790 and Murdoch and his mother-in-law were left to look after his two sons.

In the early 1790s Murdoch began experimenting with the uses of coal and took out a patent for 'the art and method of making... dyes, paints and colours from coal', the fore-runner of our modern aniline dyes. Following on from this he used coal-gas for lighting and in 1792 his house in Cross Street was the first in the world to be lit by coal-gas.

Murdoch received little praise or encouragement from his employers for his inventions but he continued experimenting with gas when he returned to the Soho Foundry. He died in Handsworth, Birmingham, in 1839.

His other inventions included a substitute for isinglass (a kind of gelatine), for which he was paid £2000 by the London brewers, a cast iron cement, a steam gun, pneumatic letter and package transmission (a device that was used in some shops until the 1950s), a method of compressing and moulding peat-moss, a pressed air platform lift, piped central heating and water saving ideas for canal locks.

William Oliver 1695-1764

The inventor of the famous Bath Oliver Biscuits, a native of Ludgvan and the physician for the Bath Mineral Water Hospital from 1740-61. William Oliver was a friend of Ralph Allen and mingled with the fashionable literary figures of the day. Before he died, William Oliver entrusted the biscuit recipe to his coachman, Atkins, together with £100 and 100 sacks of flour. Atkins set up shop in Green Street, Bath, but now the biscuits are made on a large scale and sold country-wide. Each biscuit is stamped with a portrait of Dr Oliver, Inventor.

Andrew Pears 1768-1845

The original and distinctive Pears Soap was invented by a Cornishman, Andrew Pears, the son of a farmer. But after completing his apprenticeship as a hairdresser, he left Mevagissey to try out his skills in London.

Here he established himself as a successful hairdresser and beauty specialist in fashionable Soho. However, he discovered that the harsh cosmetics (many containing arsenic or lead) used by his clients harmed their complexions and he set about producing his own brand of refined, delicate soap. Thus Pears Transparent Soap was launched.

The new product was an immediate success and to prevent imitation, each packet was signed with his own quill. Today, Pears Soap is a household word; it is manufactured in India, with the same recipe devised by Andrew Pears some 200 years ago.

James and Frederick Pool

Today J & F Pool Ltd perforate and expand materials such as stainless steel, copper, plastic or silver for use in many situations, from microwave cookers to airport seating. All this is a long way from the firm's beginnings when in 1848, James 'Tinman' Pool established his business at the rear of an ironmonger's shop in Hayle.

This early enterprise expanded when his sons, James and Frederick took over as partners in 1862. Working with wire screens, the brothers realized that different industries and products required different grades of mesh and shortly afterwards, James came up with J & F Pool's most important invention, the Cornish gauge – a plate with holes of gradually increasing size to enable the manufacturer to order the perforation to suit his needs. To begin with, the gauge was most useful within the mines for grading metal ores but similar machinery was later developed for stone grading in quarries. In 1920, J & F Pool patented 'Perfex', a rotary stone-screen for the quarrying industry, followed by the more refined 'Superfex' in 1927.

Over the years, the firm has expanded to produce many new products and is now Hayle's largest employer.

Rugby football

In his *Survey of Cornwall* (1602 and recently re-issued), Richard Carew outlines the rules for hurling, one form of which later developed into the game of rugby.

Forget Rugby School – rugby football began in Cornwall!

The Tangye brothers

Notably James Tangye 1825-1913 and Sir Richard Tangye 1833-1906. For most people, the name Tangye means novels of twentieth century life. But the family name first sprang into prominence over a hundred years ago in an entirely different speciality – heavy engineering. The Tangye brothers, James, Joseph, Edward, Richard and George, had no less than 26 patents between them and although much of their pioneering work took place at the Cornwall Works in Birmingham, they were born in Illogan and maintained their associations with the county.

James Tangye in his workshop

The launch of the Great Eastern *in 1858. Sir Richard Tangye said,*
'We launched the Great Eastern *and the* Great Eastern *launched us'*

From a young age, James in particular showed a flair for inventive designs, with his velocipede or 'dandy horse' and a huge water wheel for a stamping mill. Richard was the first brother to move to Birmingham, as a clerk, and in 1855 he was joined by his two brothers, George and James. Three years later the brothers combined forces to form the firm James Tangye & Brothers, Machinists. Here they invented lifting jacks, hydraulic presses, pulley systems and a steam driven locomotive that could reach a speed of 20 mph. The future of the firm was assured when in 1858 their hydraulic jack re-launched Brunel's prestigious ship after it got into difficulties. As Sir Richard said later 'We launched the *Great Eastern* and the *Great Eastern* launched us.' Another important achievement was to raise Cleopatra's Needle into place on the Thames Embankment.

Whilst James was the practical engineer, Sir Richard was the business manager. In Birmingham he gained respect as a model employer by improving the working conditions in the factory; he introduced a nine-hour day with a half-day holiday on Saturday. Active in local politics, he was knighted in 1894 for services to the community.

Both James and Sir Richard returned to Cornwall in retirement, James to Aviary Cottage, Illogan and Sir Richard to Glendorgal, Newquay.

Cornish National Tartan

Tartans are traditionally associated with Scotland but now Cornwall too can boast its own tartan design. The first Cornish National Tartan was devised in the 1950s by Ernest Morton Nance from Trevone, near Padstow. His design combines the Cornish national and the Duchy colours, black and gold with red, blue and white lines, interwoven with the flag of Saint Piran (a black square with a white cross).

William Thomas ('Tom') Teagle 1911-1989

Away from the coast and the tourist resorts, inland Cornwall is still very much an agricultural county and the firm Teagle Machinery Ltd at Blackwater, Truro, have been in the forefront of many of the county's agricultural inventions.

Tom Teagle was a genius of an inventor; few people in Cornwall have been responsible for so many inventions. His wife recalls that even on their honeymoon engineering came first and she had to stay with her sister in Kent whilst he travelled to the Midlands to buy steel.

The early inventions were directed towards the Cornish market and one of Tom Teagle's first machines in 1941, responding to the wartime emergency, was his novel potato planter, which not only planted but placed fertiliser around each potato. In 1942 he designed the world's first tractor-mounted steerage hoe, which could be steered by the operator independent of the tractor. One of his most important inventions appeared in 1949, a trailed fertiliser-spreader and this pioneering design was soon copied by other companies. Other inventions followed, transplanters, mounted broadcasters, seed drills, 'Spudnick' potato harvesters, pasture toppers, bale shredders and the versatile Teagle 50cc two-

Nowadays we are so used to hedge-cutters and strimmers that it is hard to realise how revolutionary this invention from the fertile brain of Tom Teagle appeared at the time

stroke engine. The 'Jetcut' lightweight hedgetrimmer broke new ground in the 1950s and the idea was taken up by the Japanese, who now produce a similar design.

New, bright red and yellow Teagle machines are still coming off the production line, but today's designs are aimed at the larger international markets rather than the small scale Cornish farmer.

Henry Trengrouse 1772-1854

The commemorative inscription on Henry Trengrouse's tombstone in Helston churchyard sums up his achievements – 'Henry Trengrouse…rendered most signal service to humanity, by devoting the greater portion of his life and means, to the invention and adoption of a Rocket Apparatus.' Although he spent most of his working life developing and trying to gain acceptance for his life saving rocket, he died in poverty, having spent £3000 of his personal fortune on a device that was to save thousands of lives.

Henry Trengrouse was born in Helston in 1772. He attended Helston Grammar School and left to start work as a cabinet maker. His career might have remained uneventful but for the shipwreck of the frigate *Anson* on the treacherous sands at Loe Bar on 29th December, 1807. This event changed the course of his life.

Like many young men in Helston, Henry Trengrouse rushed to the scene to offer help, but although so close to shore, many on the ship could not be reached and, despite rescue efforts, some sixty people drowned.

Trengrouse returned home in a daze and retired to bed, too ill to work. In trying to find a solution to the problem, he recalled a recent firework display and realised that a rocket might provide the answer. The metal cylinder rocket designed by Henry Trengrouse could be attached to and fired from an ordinary musket, from the cliff top, shore or the ship itself, carrying a light-weight rope to, or preferably from, the ship. Heavier ropes could then be pulled aboard and attached to the mast. To complement the rocket invention, Trengrouse also designed a bosun's chair (later called the Breeches Buoy) and he invented a cork life jacket or 'life pre-

28

Henry Trengrouse is famous in Cornwall for his invention of the rocket apparatus for life-saving, but he also invented a bosun's chair and cork lifejackets, as worn here by a lifeboat crew

server' and built a model of an unsinkable life-boat.

Unfortunately, Trengrouse had an uphill struggle in getting his inventions accepted and spent his fortune in expensive and largely fruitless journeys to London. He had to wait until 1818 before he was able to demonstrate his idea to Admiral Sir Charles Rowley. The Admiralty grudgingly placed an order for twenty sets, but then proceeded to make the device themselves and paid Trengrouse a mere £50 for his patented design. The Tsar of Russia expressed his gratitude for the lives saved along the Baltic coast by presenting Trengrouse with a diamond ring and the Society of Arts gave him a silver medal and twenty guineas in 1821. Trengrouse died exhausted and penniless in February 1854.

The rocket device, in all its stages of development, is now on display at the Helston Folk Museum.

Richard Trevithick 'The Cornish Giant', 1771-1833

Ask anyone living in Cornwall to name one Cornish inventor and the person that most will know is Richard Trevithick. His abilities and talents were scarcely acknowledged during his lifetime but now there is a commemorative statue outside Camborne library, an engraved stone at Pool, marking his place of birth, and his former home at Penponds is open for viewing. On the Saturday in April closest to the date of his birth, the people of Camborne celebrate Trevithick Day with a colourful outdoor market and working steam-powered vehicles parade through the crowded main streets.

The inventions of Richard Trevithick are almost too numerous to describe; he worked on his projects with single-minded determination but with little business sense and he never achieved the fame and financial success that he deserved. Throughout his life, Trevithick was fortunate in having the support of the women around him: his doting mother (he was the only son with five sisters) and later his loyal wife Jane, the daughter of John Harvey, his boilermaker at Hayle.

Apart from an aptitude for mental arithmetic, Richard Trevithick showed little signs of early genius when attending Camborne School. He was a regular truant who gained a reputation for performing extreme feats of strength and he left to work in a nearby mine. By 1792 Trevithick was promoted to consultant engineer at Tincroft, moving to Ding Dong Mine in 1796 where he became involved in experiments to improve the 'duty' of the steam engine (the efficiency in fuel consumption).

Following on from his first experiments, he began to develop his ideas on high-pressure steam, culminating in the first ever steam-propelled locomotive to carry passengers, from Rosewarne to Beacon Hill on Christmas Eve 1801. The event was celebrated with roast goose and is recalled in the Cornish song 'Going up Camborne Hill' and resulted in Trevithick's first patent in 1802. The Trevithick Society produced a full size working replica of the road locomotive for its 200th anniversary in 2001 (see cover).

The power of high-pressure steam locomotion was next

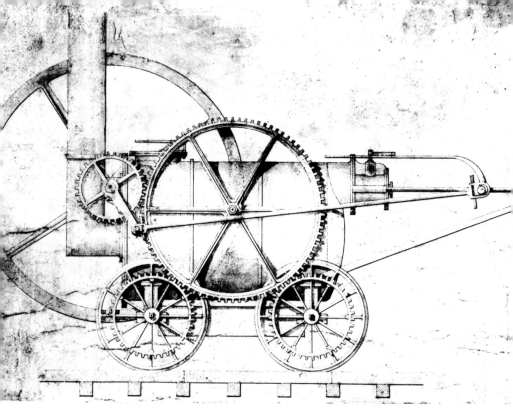

The design for another of Trevithick's locomotives, also from 1804

applied to rail transport, when in 1804 his locomotive carried ten tons of iron and 70 men for nine miles at Penydaren ironworks in South Wales.

Unfortunately, the engine proved to be too heavy for the rails which broke under the weight – but the experiment was a success and he could justly lay claim to be the first person to put a self-propelled vehicle on a railway track. George Stephenson's Rocket did not make its trial run until 1830, some 26 years later.

What was not so successful was Trevithick's attempt to use his engine as a visitor attraction on a circular track near to the present day Euston station in London, charging one shilling per person to ride on the 'Catch Me Who Can'.

Ever prepared to accept new challenges, his next contract involved working on a tunnel under the Thames and high-pressure steam was applied to nautical matters, with a steam dredger, steam tug and floating crane and a design for a screw propeller. He recognized the importance of iron in shipbuilding and suggested using iron tanks for drinking water. During his stay in London, Trevithick caught typhus and might well have died but for the caring concern of his wife, Jane, who made one of her rare excursions away from her family in Cornwall to nurse him back to health. Trevithick was declared bankrupt and returned to Cornwall aboard the Falmouth packet.

Lured by the promise of great riches, Trevithick left his family and Cornwall in 1816 and spent the next eleven years travelling around South America. No riches were to be found and Trevithick returned home penniless in 1827.

Undaunted, on his return he continued to work on new ideas – a portable room heater, a boat driven by a water jet, plans for transporting ice from the Arctic, a cast iron column in London to mark the Reform Bill and a contract for land reclamation and drainage in Holland. Trevithick's strength finally failed and he died at Dartford in 1833, aged sixty-six.

As a footnote, Trevithick's descendants live on, some of them in Japan where they work as engineers!

Charles Warrick

A late eighteenth century lawyer with singularly eccentric habits, Charles Warrick from Truro, built a boat covered with waterproof paper and attracted much attention by sailing it around Falmouth harbour. The two paddle wheels were connected by a crank, which he worked with his hands.

Chris Woolf 1947-

Proving that Cornwall is still in the forefront of new technology, Chris Woolf from Merrymeet, near Liskeard, has invented a new suspension system for microphones, whereby the suspended microphone is isolated from vibrations and extraneous noises to produce almost perfect sound quality. In March 2000 he was honoured with an Academy Award for technical merit by the American film industry.